The Challenge for Energy Justice

Raphael J Heffron

The Challenge for Energy Justice

Correcting Human Rights Abuses

palgrave
macmillan

Raphael J Heffron
CEPMLP
University of Dundee
Dundee, UK

ISBN 978-3-030-80096-3 ISBN 978-3-030-80097-0 (eBook)
https://doi.org/10.1007/978-3-030-80097-0

This Palgrave Macmillan imprint is published by the registered company Springer Nature Switzerland AG.
The registered company address is: Gewerbestrasse 11, 6330 Cham, Switzerland

purpose beyond financial gain, such as serving the public interest (Larson, 1977), yet this does not apply to the same extent to institutions. In most cases, an individual receiving a direct financial benefit represents a clear conflict of interest; yet, as Emanuel and Steiner (1995) have argued, the same standard does not apply to institutions, where financial goals are seen as necessary to achieve their "mission," which for publicly traded firms is often shareholder value (Lazonick and O'Sullivan, 2000). As the institutional pursuit of profits is not seen as self-interested behavior, "society... may erroneously be more tolerant of circumstances in which an institution's [rather than individual's] financial interests may compromise the integrity of its missions" (Emanuel and Steiner, 1995).

Institutional-level conflicts of interest are further complicated by corporate strategies that create conglomerations of diversified firms, with complex ownership structures (Davis and Craft, 2000). Corporate strategies can be implicated in encouraging the ignorance of potential harms, such as through setting high rewards for unrealistic goals (Dana, 2005) or a culture of "meeting the numbers" at any cost (Jennings, 2004).[2] As corporate strategies orient toward creating "synergies" between formerly independent entities, this may raise the level of conflict from the individual level to the institutional, creating concerns for the public interest (Davis and Craft, 2000). Codes of conduct for individuals may be incompatible with firm goals, such as in media conglomerates:

> ... the trouble is that journalism is not really conceived of as "journalism" at all, but as just another product alongside all the other products from which a diversified conglomerate expects to profit. The former seeks to reconcile the goal of serving the public with the goal of making money; the latter ignores the first goal entirely. (Davis and Craft, 2000, 223)

This can be seen as not just a conflict between the goals of the firm and that of the individual, but also between the firm and professionals charged with responsibility for the public interest. This reinforces the importance of understanding the organizational or institutional setting for planning

[2] Ironically, Jennings (2004) also argues that a culture of social responsibility, such as through philanthropy or community service, is used to overlook the violation of basic ethical principles, which is particularly concerning for planners who have long seen their work as devoted to improving the public good.

professionals, especially in the context of financialized firms focused on increasing shareholder value.

The "single-minded focus on individual acts" in professional codes of conduct has resulted in a lack of attention to conflicts that originate at the institutional level (Davis and Craft, 2000, 227). Despite the limited attention to institutional conflicts of interest, these can be particularly difficult to assess:

> Because any discretionary decision can be influenced by many factors, it may be impossible for external observers to determine what factors influenced a particular decision. Indeed, it may even be difficult for decision makers to be certain about their own motives and what factors influenced a particular decision. (Emanuel and Steiner, 1995)

Critically, the difficulty in determining institutional-level conflicts of interest is that they may exist regardless of any actions taken by individuals within the institution, such as recusal, as conflicts may arise from mergers and acquisitions or other management strategies at the level of the firm (Davis and Craft, 2000).

Professional Practices in Financialized Firms

The institutional context for planning work—or how organizational contexts impact the work of professionals in practice—takes on additional importance in light of broader economic trends of financialization. Financialization has been linked to both the emergence of finance-led economic systems, as well as the "processes and effects of the growing power of financial values and technologies on corporations, individuals and households" (French et al., 2011, 799). This trend has led to the growing influence of capital markets on a wide variety of social, political, and cultural spheres, often resulting in heightened risk, volatility, and uncertainty (Pike and Pollard, 2010). Traded firms must respond to a diverse constellation of financial actors including shareholders, fund managers, and management consultants that influence the direction and operation of firms (Williams, 2000; Froud et al., 2000). For traded firms, share price has been an increasingly important measure of firm success, solidifying the importance of analysts and investor expectations (Zorn et al., 2004). Scholarship has shown that the particular constraints of publicly traded firms, such as quarterly reporting, consistent and significant growth, and standardized performance metrics can re-orient corporate strategies toward the demands of capital markets and shareholders (Lazonick and O'Sullivan, 2000; Froud et al., 2000; Williams, 2000). These demands can be transformative for firm structures, with a shift toward chief financial officers (CFOs) to manage stock valuation, and changing acquisition strategies to meet the preferences of institutional investors (Zorn et al., 2004). The influence of financialization has been observed in the "outputs" of planning-related fields, such as the production of housing (August and Walks, 2018), infrastructure (O'Brien et al., 2018), and urban redevelopment (Weber, 2010; Rutland, 2010). Less attention has been paid to planners themselves working in traded firms, despite evidence that financialization can have a significant impact on firm employees (Cushen, 2013; Faulconbridge and Muzio, 2009).

The influence of capital markets and financial goals has been shown to impact the work of professionals in different fields, who may struggle in reconciling their professional values with the market imperatives of financialized firms (Alvehus and Spicer, 2012; Cushen, 2013). These demands are imposed on employees through mechanisms such as corporate budgeting processes, which allow financial narratives and capital interests to dominate the work of knowledge workers (Cushen, 2013). Financial demands have been observed as a form of employee control, transforming how professionals spend their time and prioritize projects (Alvehus and Spicer, 2012). Similarly, the focus on key metrics and financial measures can clash with professional values and long-held conceptualizations of professional practices (Faulconbridge and Muzio, 2009). Given the influence of shareholder value and other financial metrics in shaping the work of professionals in practice, the following sections examine how professional values are addressed by codes of conduct, and how this is negotiated through practices in publicly traded firms.

Research Methods

While planners may work for a range of firm types, this research focuses on firms broadly classified as Architecture/Engineering/Construction (A/E/C) professional service firms. Firms were selected using three criteria—(1) ranked in *Engineering News-Record* (2018, Global Design Firms by Revenue), (2) provide planning or urban design services, and (3) publicly traded—resulting in a sample of six firms (Table 1). Though

Table 1 Publicly traded firms with planning services

	ENR Ranking (2018)	*Employees*	*Number of offices*	*Headquarters*
AECOM	3	87,000	91	U.S
Arcadis	8	27,875	350	Netherlands
IBI	89	2700	60	Canada
WSP	6	32,000	500	Canada
Stantec	10	22,000	350	Canada
WS Atkins[a]/SNC-Lavalin	7	50,000	343	Canada

[a]WS Atkins (U.K.) was acquired by SNC-Lavalin in 2017

similar activities may be occurring in privately held firms, there is a methodological benefit in focusing on publicly traded firms, in that reporting requirements make financial and strategy information about firms more easily available. To understand the financial context for these firms, content analysis was conducted on publicly available documents, including quarterly reports, investor day presentations, annual reports (79 total), and employee codes of conduct (5 total). This data was triangulated through interviews with planning professionals with experience in the identified firms (31 interviews), as well as management consultants who advise the A/E/C industry (11 interviews). The interviews with planners focused on their professional experiences working in these types of firms, as well as any conflicts of interest that they perceived. Management consultant interviews covered broader financial trends affecting these types of firms and management strategies. Lastly, I analyzed the professional codes of conduct for professional planners in the U.S. and Canada to provide insight into professional values in two jurisdictions.

Findings: Corporate Strategies and Institutional Conflicts of Interest

"Protect the Interests of the Corporation": Regulating Professional Practice

Following several high-profile corruption cases involving engineering companies (Van Praet, 2016), firms have paid increasing attention to how conflicts of interest are managed, such as through employee policies, whistleblower hotlines, and integrity officers (see also Sturdy in this volume). Yet, the disconnect between firm goals and those of professional planners is perhaps most apparent in corporate codes of conduct. Corporate codes may be seen as a minimum standard, with planning professionals expected to adhere to higher expectations through their own professional licensing and codes of ethics.[3] Nonetheless, both professional and corporate codes largely display a similar orientation toward conflict of interest, with a focus on individual actions and narrow definitions of conflicts (Table 2).[4] Only two types of conflict of interest were common to all of the surveyed professional and corporate codes: accepting gifts and having a financial interest in a client, competitor, or another third party.[5] Only three firms included specific policies on public sector clients, and in most cases, this was limited to patently illegal issues such as bribery (e.g. WSP, 2019, 7).

Perhaps unsurprisingly, many corporate conflicts of interest policies are not only focused on individual behavior but also aimed at protecting the firms' financial interests. For example, in one firm code, three out of five examples of possible individual-level conflicts of interest cover issues that would impact the firms' revenue, including taking advantage of business opportunities without first offering them to the firm, investing in firm competitors, or taking outside employment (AECOM, 2018a, 20). This is also reflected in ambiguity regarding *who* would be affected by many of the issues referred to as conflicts of interest. For example, for activities such as taking advantage of business opportunities, the impacted party would be the firm, in contrast to professional codes which emphasize protection of the public interest.

[3] It should be noted that sections of the AICP Code of Conduct that refer to responsibility for the public interest are described as "aspirational" and "an allegation that we failed to achieve our aspirational principles cannot be the subject of a misconduct charge or be a cause for disciplinary action" (American Institute of Certified Planners, 2016, 1).

[4] While firms may be influenced by the regulatory context of the location where they are headquartered (i.e., Canada's foreign anti-corruption legislation is mentioned in some firm documents), all firms operate globally. In instances where firms had different codes depending on location, the global code was used.

[5] Both the Canadian Institute of Planners and the American Institute of Certified Planners provide additional guidance on dealing with ethics and conflicts of interest, such as through workbooks and toolkits. Analysis was limited to the codes for comparability with the corporate codes.

Table 2 Professional and corporate codes of conduct

Types of conflict of interest		*Professional codes*		*Corporate codes*[a]				
		Canadian Institute of Planners	*American Institute of Certified Planners*	*AECOM*	*Arcadis*	*Stantec*	*SNC-Lavalin*	*WSP*
Individual	General conflicts of interest (not detailed)	x	x					
	Accepting gifts	x	x	x	x	x	x	x
	Financial interest in competitor, supplier, or client	x	x	x	x	x	x	x
	Secondary (outside) employment or board membership		x	x		x	x	x
	Relationship with competitor, supplier, or client			x	x	x	x	x
	Relationship with another employee			x			x	x
	Taking advantage of business opportunities learned through position			x				x

Types of conflict of interest		*Professional codes*		*Corporate codes*[a]				
		Canadian Institute of Planners	*American Institute of Certified Planners*	*AECOM*	*Arcadis*	*Stantec*	*SNC-Lavalin*	*WSP*
	Property (selling and leasing to or from employer)					x		x
	Family member who is government official						x	
	Hiring former government officials						x	
Institutional	Involved in earlier parts of work (setting scope or specifications)			x				
	Self-evaluation			x				
	Unequal access to information			x[b]				
	Provide advice to public sector and other clients in same jurisdiction	x						
Other	Reference to public interest	x	x		x			x[c]

(continued)

Table 2 (continued)

Types of conflict of interest	*Professional codes*		*Corporate codes*[a]				
	Canadian Institute of Planners	*American Institute of Certified Planners*	*AECOM*	*Arcadis*	*Stantec*	*SNC-Lavalin*	*WSP*
Specific policies for public sector clients			x			x	x

[a]IBI did not have a publicly-available employee code of conduct and did not respond to email requests
[b]While listed as an organizational COI, it notes that this results in the risk of competitive advantage, not bias (AECOM, 2018a, 24)
[c]Includes reference to "respecting communities"

While all codes surveyed included various types of individual conflicts of interest, only one corporate code and one professional code covered what may be considered institutional conflicts of interest. In the corporate code that addresses institutional conflicts (referred to as "organizational conflicts of interest"), less weight is placed on addressing these types of conflicts, with little specific guidance. For most types of individual conflicts of interest, the policies require *mandatory* measures such as disclosure, recusal, or written approval from in-house counsel. However, for potential institutional conflicts of interest, the policies state "we should exercise common sense, good judgment and sound discretion to assess whether a significant potential conflict of interest exists" and work with management to resolve it (AECOM, 2018a, 24). While institutional conflicts of interest are likely more difficult to assess—and mitigation could potentially have an impact on firm revenues—these policies are much more ambiguous in how they should be dealt with.

One critical difference between professional and corporate codes relates to the "mission" of the organization. For planners, professional responsibility is first outlined in relation to the public or public interest, followed by clients, employers, and the profession (Canadian Institute of Planners, 2016; American Institute of Certified Planners, 2016). In particular, both Canadian and American professional codes note the interconnectedness of decisions and the long-term impact of decisions. In contrast, most corporate codes frame the policies in terms of loyalty to the firm and protecting its business interests. For example, one firm's conflict of interest policy begins by stating "We must ensure that we always act in the best interest of [the Corporation]" and lists the potential risks firstly as decreased shareholder value, followed by legal liability or reputational damage (SNC-Lavalin, 2020, 4.1), while others note that employees are "required to behave in a loyal manner at all times so as to protect the interests of the Corporation" (WSP, 2019, 11). One firm code does include a statement on "commitment to society," though how this is balanced with other goals such as the directive to "focus on increasing shareholder value by achieving a superior return on equity" is not addressed (Arcadis, 2018, 4). The orientation of corporate codes of conduct toward shareholder value demonstrates one way in that financial imperatives are filtered down to the level of professionals in practice, and exposes sharp differences with value systems for planners.

Corporate Strategies in Financialized Firms

While private-sector planners work in all types of firms, the rise of publicly traded mega-firms, with diverse practice areas and global scope, introduces a different set of conditions for both professional planners and their public sector clients. As noted above, the demands of financialization have impacted many sectors by increasing growth expectations, introducing fine-grained measures of financial success, and requiring standardized reporting. Engineering firms have not been immune to these trends, with the pace of acquisitions by traded firms accelerating significantly in the last 25 years, including mergers of very large firms (Linovski, 2019). Although small, specialized firms are still numerous, there has been increasing stratification in the industry, with large firms responding to the growing size, scale, and complexity of projects through widespread consolidation. As one firm noted in an investor report:

> The consulting and engineering market is one of the last professional services sectors attempting to achieve global consolidation... recognizing the need for large multinational clients in several industry sectors to work with a smaller number of larger vendors to cater to their growing global needs. (Arcadis, 2015)

As a result of this consolidation, several very large firms dominate the global engineering services market, with thousands of employees and hundreds of offices. While the long-term financial value of merging firms has been questioned (Lazonick and O'Sullivan, 2000), these aggressive strategies of merger and acquisition are often tied to increasing share value and the demands of financialized firms. As one management consultant explained, trends in consolidation can be partly attributed to the need for demonstrating growth at a level unattainable through organic means:

> When you're a publicly traded company, Wall Street demands growth. Sometimes a company can only grow 5% or 10% [through] internal, organic growth. If they make acquisitions, then they can do much more than that. And so that bumps their stock price... and I think that's kind of the norm. [Interview 40, Management Consultant]

These "mega-firms" often have significant influence in many areas of planning practice. Combined with the increasing scope and complexity of projects, this has arguably changed the nature of planning practice for many professional planners. In this context of financialization and increased consolidation, the implications can be significant, both from the professional practice perspective and for the public interest. As the following sections discuss, the financial pressures for traded firms can lead to different challenges, and potential conflicts of interest, particularly for public sector clients.

De-Risking the Firm: Corporate Strategies and Public Sector Clients

After a previous period of expansion in practice areas (Linovski, 2019), firms are increasingly choosing, or being pressured by investors, to concentrate primarily on professional services, such as architecture, planning, and engineering (e.g. Starboard Value LP, 2019). The concentration on professional services has seen a move away from more volatile private-sector activities, such as construction (Stantec, 2019) or oil and gas (AECOM, 2018b), in favor of "less risky" public sector work:

> CS [Construction Services] is largely a commercial payor business with fixed price contracts, where the Company often bears the risky contingent liability to cost overruns and project delays. By contrast DCS [Design and Consulting Services] is predominantly a professional services business serving state and federal payors that has not demonstrated the earnings volatility of the CS business. (Starboard Value LP, 2019, 8)

This increasing focus on "pure play" consulting services—in contrast to earlier periods when firms widely touted their expansion into different sectors (Linovski, 2019)—is often described in terms of de-risking the firm. The increased reliance on the public sector is evident in firms' revenue from public sector contracts, which for some firms is over 50% (Table 3; see Raco in this volume). Management consultants have noted that consolidation trends are also linked to the focus on public sector revenue:

Table 3 Firm type and revenue sources

Firm	*Pure-play*	*% of Revenue from public sector*[2]
AECOM	No[1]	51
Arcadis	Yes	55[3]
IBI	Yes	n/a
WSP	Yes	56
Stantec	Yes	39
WS Atkins/SNC-Lavalin	No	56[3]

[1]Currently in process of divesting non-consulting business lines
[2]*Sources* AECOM (2018b), Arcadis (2019), Stantec (2018), WS Atkins (2016), "WSP updates" (2020)
[3]Includes EU regulated sectors such as transport and energy

> The larger the company, it seems like the larger the clients that they need. And the larger clients with the larger projects are usually government. [Interview 40, Management Consultant][6]

While some public sector risk is acknowledged, such as with austerity measures reducing public spending (WS Atkins, 2016), firms and financial actors, such as analysts and institutional investors, predominately discuss government work as less volatile, as it is seen as less responsive to economic cycles or downturns. Public sector clients are also seen as more reliable payors, a key consideration for firms that are regularly assessed in terms of days sales outstanding (DSO), and strive to lower the time in accounts receivable (e.g. Arcadis, 2016, 9).

Despite this discourse of the safety of public sector contracts, the pressures for financial growth can impact how firms approach bidding and contracts, especially for public sector clients. The collapse of Carillion, a multi-disciplinary firm with large government sector contracts, has been attributed to its practice of bidding over-aggressively for public projects, especially in the wake of the 2008 recession when private-sector work was scarce (*The Economist*, 2018). As Cushen (2013, 327) notes, typical of financialized capitalism are unrealizable goals such as "levels of returns beyond what the product market could deliver." In addition to concerns

[6] Some planners interviewed also pointed to changes on the project side, such as larger and more complex projects, as well as increased regulatory requirements and a client preference for working with fewer vendors, as necessitating larger firms.

about the riskier construction market, financial actors have also called for firms to withdraw from business segments that involve hard-bid or fixed-price contracts (Rubin, 2019), which can introduce more risk for the firm, rather than the client, if costs are higher than anticipated. Some firms include the breakdown of their contract type in investor documents, noting that 50% of their contracts are cost-plus and emphasizing the benefits of this strategy (AECOM, 2015).

Corporate strategies such as the prioritization of public sector contracts (with accompanying pressure to obtain more contracts) and consolidation of firms—combined with the consistent pressure for increased shareholder value—create an organizational context for planners that is very different from those working in the public sector. As noted earlier, this creates conditions where the goals of the firms may be inconsistent with the goals of professionals, notably the direction to protect the public interest. The clash is most apparent in directives to sell multiple service lines to the same client, coupled with the reliance on public sector contracts and maintaining shareholder value. As discussed below, in the absence of clear institutional conflict of interest policies, there are varying strategies for acknowledging and managing these issues.

Managing Conflicting Interests

The differences between professional codes of conduct (which stress attention to the public interest) and corporate policies (which largely address firm interests) raise questions as to how they are enacted in practice. While largely not addressed by either corporate or professional codes, cross-selling strategies in firms present potential conflict of interest concerns for clients of planning work. Firms often explicitly encourage cross-selling (selling multiple services to the same client, such as planning, engineering, and project management). Despite the narrowing in firm focus, planning work can often lead to contracts in related fields, as noted by this firm executive:

> [We go] from business advisory, trying to advise them what is the best combination of solutions for the city, and then from business advisory we can drive it down into possibly programme management, project management and into design. (Arcadis, 2013, 42)

For many types of projects that planners are involved in, there are multiple work packages for each stage of work, with elements that are bid on individually but contingent on earlier scoping and assessment processes (see Fig. 1). Unlike other professions, planning often provides the foundation, or justification, for other types of work later on. For example, planners may analyze the feasibility of transportation investments by developing a business case for a new transit line or modelling future ridership in a city. The more lucrative design or engineering contracts are dependent on this early scoping or analysis work. While some planners stated that the potential for further contracts didn't impact their work and they were still able to provide independent advice, others noted the relationship between contracts:

> We [planning] are usually the first conversation with a client on a project, and if we have that conversation, if we have a good relationship, it makes it that much easier for the engineer to do the more detailed design work and then makes it even easier for the construction management team to come and do the really big heavy lifting. So, you know, the fee is bigger as we dig deeper into that. [Interview 21, Planner]

The planners interviewed varied in their approaches to cross-selling and the potential bias introduced when firms are oriented toward selling multiple services to the same clients. The majority did not believe the potential for future contracts impacted the advice they provided, though they did acknowledge corporate strategies that promoted it. Several planners acknowledged that this could create conflicts of interest but felt they were able to mitigate it through the independence of their professional expertise. The nature of unintentional bias—especially with the potential gain being for the organization, rather than individual—raises concerns about the impartiality of firms that are bidding on multiple stages of work.

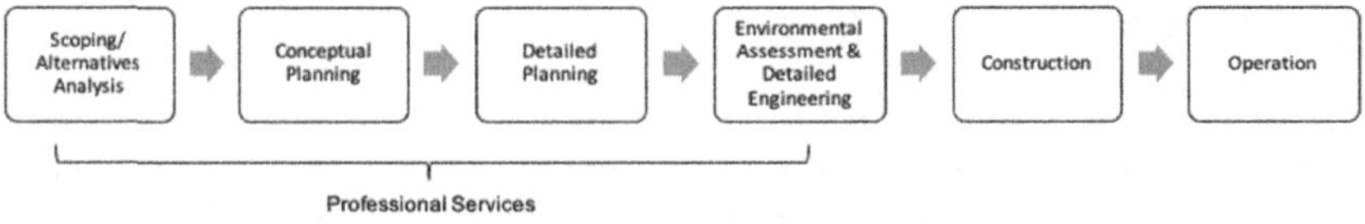

Fig. 1 Typical project cycle for planning-related work

Public sector clients differ in how they address institutional conflicts of interest, though it is most commonly managed through the procurement processes. Jurisdictions may have policies that prevent firms from bidding on work if they were involved in earlier stages, commonly referred to as "conflicted out." Agencies may also stipulate mitigation strategies, such as through disclosing previously received information to other bidders[7] or "information walls" within firms. In some instances, a conflict of interest is considered only if a firm designed the procurement criteria or is involved in firm selection (Metrolinx, 2017), despite the fact that planning services may have been involved much earlier on, such as in evaluating project options or developing a business case for alternatives. These policies also vary significantly between agencies and jurisdictions. One planner noted that while they knew of "conflicted out" policies in other jurisdictions, "but never here, we always look forward to bidding on the next phase of it..." [Interview 31, Planner]. While firms argue that the potential for future work doesn't influence their advice, involvement in multiple project stages may affect impartiality, and certainly has the appearance of conflict of interest.

While these policies are intended to limit conflicts of interest, the degree of consolidation in the sector can introduce new concerns for public sector clients. The nature and enforcement of conflict of interest policies can vary significantly, but when they are strict it can result in few eligible bidders, such as with large infrastructure or transportation projects that require significant firm capabilities. As one planner who had left private practice for the public sector remarked, policies designed to limit conflicts of interest can have unintended consequences:

> [On a project] we need to have a program management firm, an environmental or planning contract, a construction management contract, a design contract. So, how many of these big gigantic firms are there now with consolidation? It means basically that each one of these firms is going to get slotted into one of these and that's it? So, we won't have much competition? It is really sort of worrying us a little bit. [Interview 7, Planner]

[7] This is more accurately an unfair advantage rather than conflict of interest.

While this type of issue goes beyond most conceptualizations of a conflict of interest, the consolidation of firms has implications for a broader idea of the public interest:

> It seems like there are now almost single bidders for some [projects] because when the projects are so large there are only certain firms that can bid for it... And I'm not surprised that when you look at [projects] and their budgets are blown out of the water. It's not just 10 or 15% ... it's 40% more. But the chickens never seem to come home to roost, never... Taxpayers ultimately are responsible. [Interview 12, Planner]

These are institutional conflicts of interest for public sector clients—as impartiality may be compromised when the same firm is both advising on scoping or alternatives, and bidding on resulting work packages—yet there are few mechanisms to address this. Agency policies designed to limit the bias introduced by firms bidding on multiple project stages can restrict the number of bidders to the extent that new concerns about single bidders are introduced. Outside of anti-trust legislation, the management of potential conflicts has fallen to municipalities and other public agencies through procurement policies, and to firms themselves.

Conclusions: Conflict of Interest and Market Logics

While codes of conduct for planning professionals stress attention to the public interest, firms have no such obligations. At the most basic level, publicly traded firms have a responsibility to shareholders, which may be at odds with a broader public interest. As a profession that must negotiate these often-contradictory goals, planning offers a critical lens for understanding conflicts of interest introduced by private-sector involvement in public policy-making and how these may vary by institutional contexts. While the inclusion of social responsibility in professional codes does not ensure that this is adhered to, the contrast with corporate codes demonstrates a sharp difference in the expectations for professionals in practice.

Understanding conflict of interest for planning professionals working in publicly traded firms requires the reconceptualization of two common assumptions. Firstly, the idea that conflicts of interest and impaired objectivity only emerge through *intentional* corruption is countered by significant research that shows much bias is unintentional or subconscious (Cain and Detsky, 2008; Dana, 2005). Despite this, professional and corporate codes primarily operate from the assumption that conflicts of interest are not only clearly defined, but can be relatively easily mitigated.

Secondly, given the often unintentional nature of bias, assessments of financial or personal gain need to be reoriented for institutional contexts. While traditional approaches to conflict of interest assume an individual trying to get personal gain at the expense of the larger system of which they are a part of, the reverse may be more applicable in many contexts:

> As the popular definition of conflict of interest now stands, it fails to account for the reversal of the formula: that the system might act in ways directed at maximum system gain to the detriment of the individuals in the employ of the system. (Davis and Craft, 2000, 228)

For financialized firms, the idea of "system gain" exposes conflicting goals between professional practice, the public interest, and shareholder value. Although these may be regulated by public sector clients through conflict of interest policies, they are arguably ineffectual at protecting the public interest given the degree of firm consolidation and corporate strategies. Critically, since mitigating institutional conflicts of interest would likely have an impact on firm revenues (as firms have a financial interest in the outcome), it is very difficult to see how these decisions can be made by the firm, especially in the absence of clear external guidelines. While professional codes of conduct have historically focused on the individual planner, the significant changes in both the institutional context for many planners and shifting societal conditions require an acknowledgment of the broader political economy surrounding the work of planners. This will make it possible to apply a conflict of interest definition that acknowledges the role of not only individuals but institutions as well.

References

AECOM (2015) *Edited transcript ACM—AECOM investor day*, http://investors.aecom.com/Mobile.view?c=131318&d=4&v=100, accessed 15 May 2016.

AECOM (2018a) *AECOM code of conduct*, https://www.aecom.com/wp-content/uploads/2018/09/AECOM-Code-of-conduct_en.pdf, accessed 23 May 2020.

AECOM (2018b) *May investor presentation*, http://investors.aecom.com/Mobile.view?c=131318&d=4&v=100, accessed 29 May 2018.

Alvehus, J. and Spicer, A. (2012) 'Financialization as a strategy of workplace control in professional service firms', *Critical Perspectives on Accounting* 23(7): 497–510.

American Institute of Certified Planners (2016) *AICP code of ethics and professional conduct*, https://www.planning.org/media/document/9121296/, accessed May 1, 2019.

Arcadis (2013) *Capital markets day*, www.arcadis.com/investors, accessed 21 October 2016.

Arcadis (2015) *Investor factsheet*, https://www.arcadis.com/media/D/D/A/%7BDDAE6988-FFDF-49D4-BE4D-C913D2A93A53%7DInvestor%20Factsheet_web.pdf, accessed 21 October 2016.

Arcadis (2016) *Third quarter trading update*, https://www.arcadis.com/media/B/E/1/%7BBE1952C0-4200-4D70-A18A-7CDD75FFB11A%7DArcadis%20Q3%202016%20trading%20update%20-%20Analyst%20presentation.pdf, accessed 15 September 2020.

Arcadis (2018) *Arcadis general business principles*, https://www.arcadis.com/media/2/D/0/%7B2D0B1B00-9EFE-4278-B813-A4CA9B7E1112%7DArcadisGeneralBusinessPrinciples.pdf, accessed 20 May 2020.

Arcadis (2019) *Arcadis 2019 Annual Integrated Report*, https://www.arcadis.com/media/C/9/3/%7BC9319928-F9D5-4DB4-B70E-09359714C2C9%7D3Arcadis%20Integrated%20Annual%20Report%202019.pdf.

August, M. and Walks, A. (2018) 'Gentrification, suburban decline, and the financialization of multi-family rental housing: The case of Toronto', *Geoforum* 89: 124–136.

Cain, D.M. and Detsky, A.S. (2008) 'Everyone's a little bit biased (even physicians)', *JAMA* 299(24): 2893–2895.

Campbell, H. and Marshall, R. (2002) 'Utilitarianism's bad breath? A re-evaluation of the public interest justification for planning', *Planning Theory* 1(2): 163–187.

Canadian Institute of Planners (2016) *Code of professional conduct*, https://www.cip-icu.ca/Careers-in-Planning/Codes-of-Professional-Conduct, accessed 1 December 2016.

The Economist (2018) 'Cleaned out: The collapse of Carillion', 18 January.

Cushen, J. (2013) 'Financialization in the workplace: Hegemonic narratives, performative interventions and the angry knowledge worker', *Accounting, Organizations and Society* 38(4): 314–331.

Dana, J. (2005) 'Conflicts of interest and strategic ignorance of harm', in D.M. Cain, D.A. Moore, G. Loewenstein and M.H. Bazerman (eds.) *Conflicts of Interest: Challenges and Solutions in Business, Law, Medicine, and Public Policy*, Cambridge: Cambridge University Press, pp. 206–223.

Dana, J. (2009) 'How psychological research can inform policies for dealing with conflicts of interest in medicine', in B. Lo and M.J. Field (eds.) *Conflict of Interest in Medical Research, Education and Practice*, Washington, DC: National Academies Press, pp. 358–374.

Davis, C. and Craft, S. (2000) 'New media synergy: Emergence of institutional conflicts of interest', *Journal of Mass Media Ethics* 15(4): 219–231.

Davis, M. and Johnston, J. (2009) 'Conflict of interest in four professions: A comparative analysis', in B. Lo and M.J. Field (eds.) *Conflict of Interest in Medical Research, Education and Practice*, Washington, DC: National Academies Press, pp. 302–357.

Emanuel, E.J. and Steiner, D. (1995) 'Institutional conflict of interest', *New England Journal of Medicine* 332(4): 262–268.

Faulconbridge, J.R. and Muzio, D. (2009) 'The financialization of large law firms: Situated discourses and practices of reorganization', *Journal of Economic Geography* 9(5): 641–661.

French, S., Leyshon, A. and Wainwright, T. (2011) 'Financializing space, spacing financialization', *Progress in Human Geography* 35(6): 798–819.

Froud, J., Haslam, C., Johal, S. and Williams, K. (2000) 'Shareholder value and financialization: Consultancy promises, management moves', *Economy and Society* 29(1): 80–110.

Howe, E. (1980) 'Role choices of urban planners', *Journal of the American Planning Association* 46(4): 398–409.

Jennings, M.M. (2004) 'Preventing organizational ethical collapse', *The Journal of Government Financial Management* 53(1): 12–19.

Johnson, B.J. (2010) 'City planners and public service motivation', *Planning Practice & Research* 25(5): 563–586.

Johnson, B.J. (2014) 'Codes of ethics, public values and what public servants offer the bureacratic compact', *International Journal of Organization Theory and Behaviour* 17(4): 459–497.

Larson, M.S. (1977) *The Rise of Professionalism: A Sociological Analysis*, Berkeley: University of California Press.

Lauria, M. and Long, M.F. (2019) 'Ethical dilemmas in professional planning practice in the United States', *Journal of the American Planning Association* 85(4): 393–404.

Lazonick, W. and O'Sullivan, M. (2000) 'Maximizing shareholder value: A new ideology for corporate governance', *Economy and Society* 29(1): 13–35.

Linovski, O. (2017) 'Pro bono practices and government agencies', *Journal of the American Planning Association* 83(2): 180–182.

Linovski, O. (2019) 'Shareholder as client: Firms, markets, and the new business of planning', *Journal of Urban Affairs* 41(7): 1017–1035.

Lo, B. and Field, M.J. (2009) *Conflict of Interest in Medical Research, Education and Practice*, Washington, DC: National Academies Press.

Loh, C.G. and Arroyo, R.L. (2017) 'Special ethical considerations for planners in private practice', *Journal of the American Planning Association* 83(2): 168–179.

Marcuse, P. (1976) 'Professional ethics and beyond: Values in planning', *Journal of the American Institute of Planners* 42(3): 264–274.

Metrolinx (2017) *Vendor's conflict of interest policy*, http://www.metrolinx.com/tenders/en/Vendors-Conflict-of-Interest-Policy.pdf, accessed May 2, 2020.

O'Brien, P., Pike, A. and Tomaney, J. (2018) 'Governing the 'ungovernable'? Financialisation and the governance of transport infrastructure in the London 'global city-region'', *Progress in Planning* 132(100422): 1–33.

Pike, A. and Pollard, J. (2010) 'Economic geographies of financialization', *Economic Geography* 86(1): 29–51.

Raco, M. (2018) 'Private consultants, planning reform and the marketisation of local government finance', in J. Ferm and J. Tomaney (eds.) *Planning Practice: Critical Perspectives from the UK*, London: Routledge, pp. 123–137.

Read, D.C. and Leland, S.M. (2011) 'Does sector matter? An analysis of planners' attitudes regarding politics and competing interests in the planning process', *American Review of Public Administration* 41(6): 639–653.

Rubin, D.K. (2019) *Private equity pays $2.4B for AECOM government unit.* Engineering News-Record, https://www.enr.com/articles/47825-private-equity-pays-24b-for-aecom-government-unit, accessed 1 December 2019.

Rutland, T. (2010) 'The financialization of urban redevelopment', *Geography Compass* 4(8): 1167–1178.

Slade, J., Tait, M. and Inch, A. (2021) 'We need to put what we do in my dad's language, in pounds, shillings and pence': Commercialisation and the reshaping of public-sector planning in England. *Urban Studies*, https://doi.org/10.1177/0042098021989953.

SNC-Lavalin (2020) *Code of conduct*, https://www.snclavalin.com/~/media/Files/S/SNC-Lavalin/documents/policies/1003_en.pdf, accessed 24 May 2020.

Stantec (2018) *Stantec 2018 Annual Report* 132.

Stantec (2019) *Stantec announces fourth quarter and full year 2018 results and dividend increase*, https://www.stantec.com/en/news/2019/stantec-fourth-quarter-2018-year-end-results-and-dividend-increase, accessed 15 September 2020.

Starboard Value LP (2019) *Letter to Michael S. Burke, CEO, AECOM*, http://www.starboardvalue.com/wp-content/uploads/Starboard_Value_LP_Letter_to_ACM_CEO_and_Board_06.20.2019.pdf, accessed 1 July 2019.

Steele, W. (2009) 'Australian urban planners: Hybrid roles and professional dilemmas?', *Urban Policy and Research* 27(2): 189–203.

Thomas, J.M. (2019) 'Socially responsible practice: The battle to reshape the American Institute of Planners', *Journal of Planning History* 18(4): 258–281.

Van Praet, N. (2016) 'SNC looks to 'turn the page' in Quebec with repayment deal', *The Globe and Mail*, 10 May.

Wachs, M. (ed.) (1985) *Ethics in Planning*, New Brunswick, NJ: Center for Urban Policy Research.

Wargent, M., Parker, G. and Street, E. (2020) 'Public-private entanglements: Consultant use by local planning authorities in England', *European Planning Studies* 28(1): 192–210.

Weber, R. (2010) 'Selling city futures: The financialization of urban redevelopment policy', *Economic Geography* 86(3): 251–274.

Williams, K. (2000) 'From shareholder value to present-day capitalism', *Economy and Society* 29(1): 1–12.

WS Atkins (2016) *Annual Report*, https://investors.atkinsglobal.com/~/media/Files/A/Atkins-IR/reports-and-presentations/full/ar-2016.pdf, accessed 24 February 2017.

WSP (2019) *Code of conduct*, https://www.wsp.com/en-CA/who-we-are/corporate-responsibility/ethics-integrity, accessed 23 May 2020.

WSP (2020, April 21) Updates on business continuity; withdraws financial outlook. *Environmental Analyst*, https://environment-analyst.com/global/105498/wsp-updates-on-business-continuity-withdraws-financial-outlook.

Zorn, D., Dobbin, F., Dierkes, J. and Kwok, M.S. (2004) 'Managing investors: How financial markets reshaped the American firm', in K. Knorr Cetina and A. Preda (eds.) *The Sociology of Financial Markets*, New York: Oxford University Press, pp. 269–289.

The Governance of Management Consultancy Use: Practices, Problems, and Possibilities

Andrew Sturdy

Introduction—The Problems with Management Consultancy in Public Services

The provision of expert advice in government and its potential threat to democracy is not a new phenomenon (Grundmann, 2018), whether as technocracy or through experts acting as mere "servants of power" (Brint, 1990). In recent decades, policy advice has been increasingly externalized and commercialized to diverse professional service firms (PSF), notably, and our focus here, management consultancy firms and divisions of other PSFs such as the "Big Four" (Craft and Howlett, 2013). This process has been attractive to some policy actors for the relative speed of delivery ("fast policy"), the offer of an "outsider" or "modernizing" (e.g. management-based) view and the greater opportunity, as clients, to shape the content of policy directly. However, critiques emerged in the 1990s/2000s, especially around what was termed *consultocracy*—how public scrutiny and/or democratic processes were being by-passed in favor of consultants and their market-based approach (Martin, 1998). These and other concerns continue today. This chapter seeks to examine

A. Sturdy (✉)
University of Bristol, Bristol, UK
e-mail: andrew.sturdy@bristol.ac.uk

C. Hurl and A. Vogelpohl (eds.), *Professional Service Firms and Politics in a Global Era*, https://doi.org/10.1007/978-3-030-72128-2_16

how they have been addressed by outlining the governance of consultancy use and its limitations and begins to explore some of the possibilities for improvement.

Initially, criticisms of consultancy use in governments were rather tentative—consultants were mere "footsoldiers" (Hodge and Bowman, 2006) and growing usage did not necessarily equate with increased influence (Saint-Martin, 2000). Moreover, consultancy use by governments was not as widespread then, but concentrated in anglophone policy advisory systems (Craft and Halligan, 2020). More recently, use of both policy and management consultancy has increased and spread geographically, in part with the rise of e-government and international development agendas (Saint-Martin, 2012; FEACO, 2018; Steiner et al., 2018) and with the emergence of new relationships and diverse policy actors and networks such as those in transnational spaces (see OECD, 2017). Research also now seems less equivocal in outlining negative "democratic" outcomes, not least given the size and power of leading professional service firms where regulation is resisted and "revolving doors" blur the boundaries of public and private actors and interests (Raudla, 2013; O'Mahoney and Sturdy, 2016; Morgan et al., 2019). Ylönen and Kuusela (2019) for example, see *consultocracy* as leading to:

- The appropriation, privatization, and erosion of public sector knowledge, creating client dependency and over-use of consultancy.
- A weakening of public accountability/scrutiny behind commercial confidentiality.
- Privileging instrumental rationality (e.g. efficiency) over values-based criteria and alternative logics (e.g. public service ethos).

Similarly, research by Vogelpohl (2018) showed how the logic and practice of political participation are quite alien to large consulting firms such that, even when seeking to elicit public engagement in policy, they ended up "muting citizens' voices" (cf. Stapper et al., 2020).

Such critiques focus mainly on the *democratic* implications of consultancy use in public services, but it could be argued that the *potential* value of consultancy lies elsewhere, especially in creating financial efficiencies and organizational effectiveness (Kipping and Saint-Martin, 2005). However, recent research has also cast doubt on the efficiency of consultancy usage in the public sector (Kirkpatrick et al., 2018). Indeed, the industry and its users have long experienced sustained academic and public criticism across all sectors on various counts of incompetence or wrongdoing, just as its use by willing clients has grown (Sturdy, 2009). These criticisms can be split between questions over management *and* consultancy—content and process. The latter derive from the combination of consultancy being a *service* whose nature and outcomes are typically highly ambiguous and having a "commodified" and often highly valued form. This can intensify both commercial interests (e.g. fee income over client/societal need) and the incentive for both parties to assert success and hide failure because they cannot easily be established objectively (Sturdy, 2009). There is for example, little incentive for consultants to walk away from a project or "speak truth to power," even in the most non-legitimate or authoritarian political regimes (see Jones, 2019). This lack of transparency also means that, overall, consultancy can only be partially governed by the market or competition as reputations often seem to be resilient.

Given the duration and extent of criticism of management consultancy and its use, it is remarkable that there has been such little research attention directed towards reform whether targeting clients, consultants, intermediaries, or organizations and individuals. Some policy options are implied, such as non-use of consultancy or promoting the deployment of alternative sources of expertise. Also, and as we shall see, governing consultancy use through "professional client" purchasing has been a sustained field of research. However, there has been no systematic attempt in research or more generally to bring together and explore diverse options at governance in its broadest sense, including regulation and other "*systems of steering society*" (Kourula et al., 2019, 1104). Such a task is important in a number of respects. Firstly, if external management consultancy use poses a threat to democracy and/or public service effectiveness, alternative approaches need to be considered. Secondly, on the assumption that there will sometimes be value in (and demand for) using external sources of expertise, how this is best governed needs to be identified. Indeed, the critical nature of much academic research toward *management* means any real or potential benefits of *consultancy* are often effectively silenced. One exception to this universalism is Ylönen and Kuusela (2019, 15) who conclude by asking "how could the public sector itself internalize some of the good things sometimes associated with consultants (providing alternatives, agility, and greater interaction) without losing its ethos?".

Of course, outside of the consultancy context, research has explored similar issues, often under the labels of opportunism, professional or organizational wrongdoing or unethical behavior (Palmer et al., 2016). A long-standing, but still contemporary and highly relevant example is the regulation of the big (four) accountancy (audit) firms. They have been associated with various scandals as well as more habitual malpractice such as supporting tax avoidance or opportunism in their related consulting roles, which often exceeded audit in terms of revenue generation. Here too, financial pressures are seen to be of central importance rather than rogue individuals or "bad apples." As Brooks (2019, 277) notes: "good intentions alone will never be able to counter the commercial forces (incentives) that shape their (accountants') thinking and actions." In other words, the origins are seen to be systemic—wrongdoing as normal (Palmer, 2012). They lie either in the structuring of professions (and clients) such that self-interest prevails ("bad barrels") or changing institutional boundary conditions that give rise to new opportunities and incentives (e.g. regulatory voids in transnational spaces) ("bad cellars") (Muzio et al., 2016). If this is the case in a relatively clear and established knowledge domain (auditing), subject to both state and professional/self-regulation, it has sobering implications for day-to-day management consultancy practice where, as we shall see, such conditions do not generally apply.

Given the extent of wrongdoing in a *regulated* and professional field such as accounting, it seems clear that introducing formal governance *on its own*, or even culture-based initiatives (e.g. ethics training), would not be sufficient to address concerns over consultancy use. Rather, a wider—governance—view of the issues and options is needed. This is also a question of values and politics (Haines, 2013) and of helping to create the *conditions of possibility* for better practices. The remainder of the chapter seeks to begin to engage with this agenda to help stimulate debate and is structured in the following way. Firstly, existing direct and indirect approaches to the governance of consulting and its use are outlined before identifying some of the problems they give rise to. This then informs a discussion of emerging governance prescriptions, including changing the broader context and content of management consultancy, before concluding with a call for further research.

Existing Governance

Indirect Regulation

As we shall see, management consultancy and its use are often considered to be very weakly regulated, whether by the market, profession, or state. While obviously subject to contract, employment and/or company law like any other service provider, even here, the ambiguity of consultancy outcomes has led to calls to strengthen the power of courts over legal liability for firms' advice (*New York Times*, 2013; also McKenna, 2007). While weakly regulated itself, there is in fact, a strong relationship between consultancy and regulations—an *indirect* one. Indeed, management consultancy was partly founded and developed on the back of the regulation of others. In particular, McKenna (2006) charts historically how strategy consultancy in the USA emerged and prospered opportunistically. It moved into regulatory gaps created by anti-monopoly legislation and decrees such as banning banks from reviewing client management in the 1930s and prohibiting IT firms (IBM) and notably, audit firms from consulting in the 1950s and 2000s, respectively (see Kipping in this volume). Such indirect effects continue to be important. In particular, ongoing regulation of the Big Four auditing firms has had some clear impacts on consultancy. While many critics rightly consider the regulation so far as relatively toothless (CRC, 2014; Brooks, 2019) and firms' claims of "Chinese walls" between advisory (consultancy) and audit to be questionable (Mitchell and Sikka, 2011), it can sometimes have significant outcomes. Not only does it draw media and activist attention to the issue of conflicting interests, but there is evidence of changes in consulting practice. For example, the 2013 Dutch Audit Profession Act, which was stricter than the related EU regulation, led to significant restrictions or "collateral damage" on consultancy work when firms had to move to new audit clients where consultancy work had been underway (Taminiau and Heusinkveld, 2019).

State Certification and Professional and Organizational Codes of Conduct

While regulatory concerns with audit may sometimes encroach on consulting practice, more generally and directly, the regulation of consultancy contrasts sharply with that of the professions that the industry has mimicked or allied itself to such as accountancy, the law, and engineering (Muzio et al., 2011a; David et al., 2013). As is often claimed, it is an occupation that anyone can enter, without any registration or qualifications—only in Austria is there a state requirement for certification (Gross and Kieser, 2006). Of course, there is some *self*-regulation, but not to the extent of those occupations where members have the (often albeit limited) threat of being prevented from practicing or being held to account for their actions (Kipping and Saint-Martin, 2005). Rather, it comes in the form of industry, firm, and/or professional associations' *codes of conduct* (e.g. FEACO; MCA), including the ISO standard 20700 (ISO, 2017). These are often supported by training and sometimes form part of different national certification systems such as that linked to the Certified Management Consultant (CMC) designation which is recognized in 50 countries (see Linovski in this volume). They may require evidence of professional practice, but together cover a very small proportion of practicing consultants (Collins and Butler, 2020). Rather, consultancy is mostly considered as a failed or "hollow" (Kipping, 2011) profession in a traditional sense or as a "corporate profession" (Muzio et al., 2011b), including some informal closure among those in elite PSFs (Kipping et al., 2019). Indeed, the big employers, to a greater or lesser extent, cover ethical issues through training and performance evaluation in order to encourage ethical practice and help safeguard their corporate reputations (see O'Mahoney, 2011).

Purchasing Regulation and Guidelines

The regulation and wider governance of consulting use are most visible, both in practice and research studies, in the field of purchasing or procurement of services where "professional client" practice has developed or been imposed over recent years. Here, we find some hard regulation in the form of legislation on public procurement such as in the EU and nation-states (e.g. the Netherlands, Sweden), derived originally from the WTO (1979/81) (Lindberg and Furusten, 2005). As a result, there has been a shift away from the traditional "relational" or collaborative contracting, based on informal mechanisms and trust within a client-consultant dyad, to a more formalized transactional, "openly" competitive approach around a client-consultant-purchaser triad (Lonsdale et al., 2017). This can take various forms organizationally (Patrucco et al., 2018) and culturally (Pemer et al., 2014), but common features are the requirement for clients to demonstrate (formal) rationality through completing "business cases" for projects over a certain financial value.

Here, they typically answer questions on why *internal* resources are not to be used and the project's scope and measurable outcomes. This process begins to address one of the most important challenges in governing consulting—its ambiguity. In particular, bids need to be in a comparable (i.e. standardized) form for selection decisions to be made, although, as we shall see, this can generate new problems (O'Mahoney et al., 2013). Indeed, the use of Preferred Supplier Lists and framework agreements, rather than open tender, is common to minimize the (transaction) costs for both parties and in recognition of the value of trusted relationships, which they serve to embed (Sieweke et al., 2012).

Such purchasing practices and regulations are reflected in client training and development programs as well as the rise of the purchasing profession which typically challenges the traditional view that consultancy cannot be effectively evaluated (Werr and Pemer, 2007). Also, they are evident in generic guidelines issued by diverse bodies such as the OECD's (2009) reports on public procurement and monitoring for accountability and on national policy advisory systems governance (OECD, 2017; see also Institute for Government, 2019). Likewise, guidelines might be specifically directed toward the use of consultants. In the UK for example, there have been numerous studies and reports on government use of consultancy by the National Audit Office (NAO, 2016) as well as other bodies (e.g. Cabinet Office). These typically examine and prescribe "best practice" for government departments in terms of *monitoring, internalizing, and contracting*, but are not binding:

- *Monitoring*—evaluate future consulting needs annually (e.g. business cases); accurately categorize and record spending; measure project outcomes; report method of spending reduction.
- *Internalizing*—measure and develop/recruit/share skills and staff internally (e.g. project management) according to planned need (skills gap analysis/HR planning); share supplier performance information internally (also Mohe, 2006).
- *Contracting*—reward more on fixed price/incentive-based against measurable objectives (also common to recommend prioritizing knowledge transfer; robust competitive tendering/framework/approval process; integrate with supplier performance data; early legitimate engagement with suppliers; ensure staff are skilled and incentivized to manage projects effectively) (NAO, 2006, 2016).

At heart, they reflect an almost reluctant acceptance that external consultancy will sometimes be needed, but that its use has been excessive because of incompetence or informal client motives to legitimate change and consultancy pressures to sell beyond client need. Kipping and Saint-Martin (2005, 461) described such soft regulation as evidence that any government "addiction" to consultancy was, at least partially, "controlled."

Business Appointment Rules

Such reports from governing bodies are aimed largely at achieving *value for money*, while purchasing guidelines and legislation focus on supporting *competition*. This is also important in a related area of governance, but one also directed toward preserving *democratic* practices—the "business appointment rules" in employment codes and contracts designed to reduce the potentially corrupting effects of "revolving door" appointments of politicians and civil servants into consulting (and/or lobbying roles). For example, rules may impose a time limit before which a former civil servant or politician may take up a post with a service provider to government (Tyllstrom, 2019). This problem and the related issue of former consultancy executives moving into government or regulatory roles is particularly problematic and evident in the Big Four and other large consulting firms, given their pervasiveness as advisers to governments and their financial resources and skill in non-transparent, often informal lobbying (Brooks, 2019; Cave and Rowell, 2014). However, and as we shall see more generally, such regulations are often very weakly enforced (NAO, 2017; House of Commons, 2019).

Choosing Alternatives

Where contemporary governance is perhaps more effective is outside of formal ("hard" or "soft") regulation, in the discretion of actors to ban, limit or simply not to use consultants. This is not often discussed or researched (it is hard to research an absence), but is occasionally mentioned as occurring at national and organizational levels. Internationally for example, consultancy use is highly concentrated among a few countries, and not just the richest (Sturdy and O'Mahoney, 2018). The same applies to particular sectors, including public and private distinctions. In Europe for example, public sector use varies from 7 to

36% of each country's total, including variation among high consultancy users in general, such as Germany (7% public sector use) and the UK (21%) who together make up almost 60% of European management consultancy (FEACO, 2018). Likewise, some governments (e.g. China) have proscribed the use of overseas consultants (O'Mahoney and Sturdy, 2016). Similarly, partly in response to adverse media attention and popular opinion, political parties have called for bans or limits, both when in opposition and, less frequently, in government (Weiss, 2019). At organizational levels, strict annual budgets on consulting use can be applied (Mohe, 2005) and expenditure can also vary in line with senior minister or civil servant tenure and electoral cycles for example (Pemer et al., 2020). Even outright bans sometimes occur, often based on the preferences of CEOs (Faust and Schneider, 2014). Indeed, such practices suggest that reputation can sometimes serve as a governance mechanism for the industry as a whole (Glückler and Armbrüster, 2003; Karpoff, 2012), but also for individual firms who spend considerable resources on developing and maintaining their credibility (Harvey et al., 2017). Finally, and as we shall discuss below, a more sustained, but still vulnerable, approach to govern external consultancy use is reflected in the development, use, and rise of alternative sources of expertise. This is especially evident in diverse *internal* management consultancy units and through various communities of practice, think tanks, and trade/sector associations (OECD, 2017). For example, it is reported that 75% of German publicly traded companies have some form of internal consultancy operation (Mohe, 2005).

Limitations of Existing Governance

The continuing, even escalating, critiques of consultancy use in government noted earlier suggest that existing forms of governance are not being implemented effectively and/or are not sufficient. For example, in the UK at least, periodic reports by agencies such as the NAO or Cabinet Office on the use of consultancy make very similar recommendations to improve practice each time. The implied lack of progress is sometimes made explicit. For example, a 2016 NAO report looked at the extent of change since a 2010 Public Accounts Committee study on the same subject (see also NAO 2006 for a similar retrospective). This is confirmed elsewhere with the view that the UK "civil service has displayed long-standing weaknesses in how it uses specialist skills" (Guerin et al., 2018, 5). Indeed, the failure of clients to adopt systematic or formally rational approaches to consultant selection, contracting, and evaluation has long been reported more generally (Roodhooft and van den Abbeele, 2006) (Paradoxically, consulting firms evaluate both projects and consultants intensively, but these are rarely visible externally—Sturdy et al., 2009). Personal or network contacts and perceived reputation and experience still dominate selection preferences (Glückler and Armbrüster, 2003). Even where systematic procedures exist, implementation is partial. In the UK NHS for example, proposed projects over £50,000 in value have to have a business case approved by the regulator (NHSI), but there is little evidence that it assesses outcomes or whether information is shared on consultancy performance (Kirkpatrick et al., 2018).

Of course, it could be argued that such an apparent and consistent failure is merely a recognition that the requirements are *not implementable*. As noted earlier, assessing the quality of consultants and projects is often considered impossible such that informal, relational criteria need to be used instead (see below) (Wright and Kitay, 2002). Indeed, some see the failure of consultancy to professionalize as an occupation as being based, not only on the opposition of large and powerful firms (David et al., 2013), but on the nature of its knowledge base—"too elusive, fuzzy and perishable" as well as often co-produced with, and for, clients (cf. Muzio et al., 2011a, 807). Others point to cultural and structural constraints. For example, aside from the regulatory gaps created by transnational policy practice (Muzio et al., 2016), "accountability gaps" arise from the increasingly complex networks of service provision between sectors "with inconsistent arrangements for oversight, inspection, regulation and scrutiny" (Guerin et al., 2018, 4)—the "bad cellar" view mentioned earlier or what has come to be known as "new governance" (Richardson, 2012). On top of this is a more familiar, secret, and adversarial political culture which, combined with the media scrutiny, can lead to defensiveness and compliance rather than learning (Guerin et al., 2018). Likewise, on the part of consultants, there is little incentive to complain of procurement malpractice among clients, for fear of being blacklisted (Lindberg and Furusten, 2005, 182). More generally, often around the cloak of "commercial confidentiality," occupations and firms value secrecy in support of their own commercial interests and autonomy

(Vriens et al., 2018)—"bad barrels." Furthermore, the possibility of reputation or the market serving as an effective governance mechanism is limited by the ambiguity of many consulting outcomes and client dependence or lock-in as well firms' PR practices. Consider the various scandals associated with firms such as McKinsey and the Big 4 and their continuing strong reputation and use by elites and more generally (O'Mahoney and Sturdy, 2016).

The structural and cultural contexts also inform a more active failure to comply with governance—*resistance*. This has been documented in various forms, including breaking the law and helping to change it through lobbying and influencing legislators. For example, Lindberg and Furusten (2005, 177 and 182) document various ways in which clients "frequently" get around or ignore (break) procurement regulations with the overall effect that personal ties or "established business relations direct the market." One common practice is that of "*chaining*" where a series of lower value projects are commissioned which individually are not large enough to warrant a business case or going out to tender (Ylönen and Kuusela, 2019, 8). Likewise, Pemer and Skjølsvik (2018) show how, over time, both clients and consultants watered down (changed) regulations, initially by fighting regulators and then by "smuggling in" quality-based measures and selection criteria such as consultants having the "ability to connect" or "immediately create trust" (also Lonsdale et al., 2017). Such resistance is partly informed by a preference for relationship approaches to contracting, but also by pursuing personal or sectional interests over that of the organization. In particular, there is little incentive to evaluate projects when there is a risk of exposing personal failure (Honer and Mohe, 2009; Werr and Styhre, 2002). At the same time, of course, firms also act to secure their own interests, as has been evident in the resistance to breaking up the audit-consulting link in the Big Four (*Financial Times*, 2019) and to efforts to make lobbying more transparent (Cave and Rowell, 2014). Indeed, large consulting firms (along with the state) have been ambivalent at best and often practiced "outright resistance" toward regulation through professionalization (Muzio et al., 2011a, 817).

The failure of established forms of governance is not solely a matter of technical flaws, cultural contexts, and active resistance. It is also likely to be a consequence of *not fully addressing the key challenges*. Current mechanisms are partial or insufficient, not least for being directed mostly at securing "free" competition or value for money rather than democracy or managerialism and instrumental rationality. Furthermore, as O'Mahoney points out, many of the existing practices directed at rendering management consultancy ethical are highly individualized—the "bad apple" position. In other words, a "lack of institutional accountability places the individual as the arbitrator of risk: it is they, rather than the company or the government, that becomes responsible for ethical behaviour" (2011, 110; cf. Shaw, 2019). This echoes concerns beyond the consulting context, where calls for individuals to deploy cognitive or emotional methods of self-awareness—"gut checks"—are deemed useful, but insufficient (Palmer, 2012). Finally, and in parallel with debates in the field of accounting, the *financial, rhetorical, and network* power of some of the global firms create a power imbalance over clients in many contexts which warrants additional governance (O'Mahoney and Sturdy, 2016; Boussebaa and Faulconbridge, 2019). Overall then, it seems clear that not only do existing governance mechanisms need to be evaluated and, if appropriate, strengthened or better implemented and resourced, but new or adapted approaches should be considered which span a wider range of activities and issues—beyond ethics—to which the chapter now briefly turns.

Emergent Prescriptions—Hybridity and Diversity

Hybrid Regimes of Accountability, Purchasing, and Ethics

As noted earlier, despite extensive criticism from the academy and elsewhere, little research attention has been given to the governance of management consultancy. In keeping with the emphasis on purchasing-related approaches, any policy implications often relate to tightening up procurement. For example, there have been calls for early publication and scrutiny of client decision criteria on commissioning consultants and of the terms on which consultants are employed. Likewise, others have promoted the compulsory review of staff and projects against codes of conduct and/or client business cases, including knowledge transfer requirements, and requiring clients to maintain and use databases on consultancies and past projects (Brooks, 2019, 285; Guerin et al., 2018; Stapper et al., 2020; Vogelpohl, 2018; Mohe, 2006; Honer and Mohe, 2009).

Such prescriptions are helpful, but it is also important to consider possibilities within a framing of some of the wider and long-standing *dilemmas* surrounding governance in general, such as those relating to accountability, where hybrid (mixed) or alternative options appear common. For example, Vriens et al. (2018) call for what they term *conditional accountability*. They highlight the familiar tension between formal calculative rules or targets around accountability, which are often dysfunctional or counterproductive, and a softer or informal narrative approach of explaining conduct in context. Conditional accountability lies in the middle, with a focus not on conduct or results, but on conditions, which enable professional conduct. In other words, "showing that professionals have the time, tools, regulatory potential, information, or incentives, to actually and properly apply their specific knowledge and experience and dedicate themselves to realizing some societal value" (2018, 3). This is grounded in a positive, perhaps optimistic, view of the potential of professional conduct where market or bureaucratic notions such as client need or efficiency are secondary to "societal value" (cf. Freidson, 2001). However, it recognizes some of the challenges of governance—between regulation and culture for example (Palmer, 2012)—especially in ambiguous knowledge domains such as consultancy.

Similar calls for intermediate or contingent positions are evident in the consulting literature, around purchasing. For example, Lonsdale et al. (2017) recommend that the purchasing function both counters the personal engagement model and seeks to generate credibility and social capital with end-users by gaining service-specific knowledge and focusing on value for money (cf. cost) in a supporting role. This resonates with the recent recognition in operations management research of the disadvantages or "dark sides" of social capital (being too close) as well as its benefits, pointing therefore, to an optimum level in given contexts (Villena et al., 2011). Similarly, in consulting, Honer and Mohe (2009) advocate a move from *control-based measures* (e.g. purchasing rules, reporting, and centralization), which managers typically seek to bypass, to an *incentive-based* approach. Here, the purchasing function or project office is again, more service-oriented or voluntary but end-users may also have incentives to report anonymously on other clients' projects (whistleblowing) (see also Ayers and Kaplan, 2005). Indeed, others have called for the use of shaming, among peers especially, as an effective deterrent in comparison to regulation for accountability (e.g. Wedel, 2014, 273). However, this sits alongside other calls to move away from the secrecy of blame cultures to those of learning and improvement noted earlier (Guerin et al., 2018).

The dilemmas of control and incentives, regulation and self-regulation, or hard and soft law are, of course, familiar and long-standing. Likewise, one conventional resolution, albeit with its own challenges, is a more *contingent* view. This is evident in recent calls for a move away from a transactional *versus* relational purchasing dichotomy towards complementarity. For example, one might acknowledge that the outcomes of some consulting projects are easier to specify, and then manage accordingly, rather than claim either position as a universal. At a more general level, it is evident in recent calls for more integrated approaches of "strategic commissioning" and "system stewardship" (O'Flynn and Sturgess, 2019). Here, it is argued that, because public sector organizations often no longer have complete control over the whole policy process, more integrated approaches are needed with external parties. These might comprise multi-party networks which are transactional in some instances and relational with others. In the disarticulated state it is claimed, "the core work of government is stewarding these complex systems towards purpose and outcomes" (ibid., 15). Such seemingly emergent or "new governance" (Richardson, 2012) approaches of diverse and multiple actors in different forms and timescales of relationship are also evident in the private sector where the notion of "co-production" of knowledge has also become firmly established in many contexts (Heusinkveld et al., 2011) although networks are not a panacea for governance.

Finally, mixed regimes in purchasing might be matched by similar approaches to ethics such as that implied in O'Mahoney's call to de-individualize consulting ethics in formal codes, training, and business school education by emphasizing the "structural and institutional implications of ethics and not simply focus on the day-to-day micro-level issues that consultants influence" (2011, 111). Once again, this resonates with the notion of moving beyond "bad apples" to the "barrels" and "cellars" and beyond.

Alternative Actors, Organizational Structures, and Forms of Knowledge

As we have seen, given that dilemmas are inherent to governance mechanisms, there is a limit to what can be achieved through both existing and emerging regimes. However, partly echoing the conditional approach to accountability mentioned above, one can seek to change the broader *context* in which consulting occurs in order to support the actions of politicians, citizens, civil servants, and consultants in ways, which might address, mitigate, or remove some of the problems of consultancy use. This can be seen in interrelated terms of different actors becoming involved; increasing the diversity of consulting organizational forms; and changing some of the processes and forms of knowledge used in consultancy. In particular, we briefly consider the role of alternative *providers* as well as the importance of *monitors* and *activists* as actors in the consultancy field.

As is well established in debates about the role of the *Big Four* in auditing, a key problem in consultancy and professional services more widely is the concentration of service provision and the need for alternatives and greater diversity. This resonates with different concerns such as those over competition, conflicts of interest and service standardization, but mostly over power imbalance in relation to all but the largest clients. While attempts to limit the audit market of the Big Four and allow new entrants in continue to be resisted, the related global concentration of *consultancy* providers persists, but with little, if any, public debate. Notwithstanding shared features of commercial external consultancy, such as those outlined earlier (e.g. ambiguity; pressure to over-sell), *important differences between providers can also exist* in terms of culture, structure, or ethos. This might resonate with the distinction between ethics and social responsibility. For example, founders and/or particular consultants might have a commitment to a specific approach or societal goal that overrides or lessens commercial "imperatives." Such variation has been observed in the fields of CSR (Furusten et al., 2013), climate change (Keele, 2018; Owen, 2019), and politics (Stapper et al., 2020) and in seeking to develop more feminine forms of consulting (Marsh, 2009). It is also sometimes implied in client preferences such as those in the UK NHS actively choosing consulting firms with a not-for-profit structure and NHS experience (Gibbons, 2019). More generally, the diversity of organizational forms in consulting has been attributed to its relative lack of professional regulation, compared to accounting and the law (Kipping and Kirkpatrick, 2012).

If then, there are some external consulting providers who are less problematic in relation to the problems we have identified, how might they be supported? Firstly, there is some indication that conditions are already becoming favorable for greater diversity. The concentration of management consultancy—the twentieth-century model of big global players—is regarded as being under threat of "disruption" from different economic, technological, and geographical directions (Christensen et al., 2013). So-called "challenger firms," independents and networks are emerging from the "gig economy," "developing" economies and elsewhere (Meere, 2019). Such diversity could be strengthened in many ways. For example, as has been proposed in auditing, the term of client engagements could be limited in time—"*serial purchasing*" (Mohe, 2005) although this would not remove risks of consultant opportunism or managerialism.

What is needed to improve diversity are policies to support new organizational structures and processes. Here we can learn from those seeking to change capitalism more generally such as the *Corporate Reform Collective* (CRC) (2014, 180) who argue for "alternative, socially and publicly accountable corporate forms of ownership and governance." Here, given that commercial pressures have been identified as a key problem in consulting use, alternative financial models and "not for profit" ownership arrangements could be crucial. What this would look like in consulting and how it could be achieved is an area for further research, (e.g. BCorp firms) but many corporate reformers also point to the potential of engaging and empowering diverse stakeholders to be informed and apply pressure to represent more than partners, owners, or investors and include citizens and employees (Brooks, 2019; Muzio et al., 2016). Here, an obvious first focus of attention would be the large consulting firm structures such as the "*up or out*" promotion system in both professional partnerships ("P2") and private corporations (cf. Greenwood and Empson, 2003). These help generate huge pressures on securing surpluses and repeat business i.e. over-selling rather than a "culture of challenge" (Ramanna, 2019). However, as Palmer (2012, 273) points out on organizational structure reform in general, "it is difficult to align incentives so that they motivate right doing, without inadvertently motivating wrongdoing."

A longer term, wider, and more radical objective would be to seek to address the *content* of management consultancy as well as its structures and processes. This implies challenging the kind of knowledge that predominates in consulting and its wider valuation in society—a particular form of instrumental rationality and determinism (economic and technological). For example, Ylönen and Kuusela (2019, 15) argue that, so long as private-sector models are chosen or dominate thinking, "there is little hope of overcoming the problems associated with consultocracy." Such an agenda resonates with wider debates around management education and the role of the business school (Parker, 2018), but needs to encompass consulting as well (Engwall et al., 2016). At least part of this should include strengthening the presence, visibility, role, rewards, and credibility of *alternative* sources of external advice, beyond consultancy such as thinktanks, sector associations, NGOs, universities, and foundations which may draw on different knowledge bases and traditions as is evident in studies of national *policy advisory systems* (Craft and Halligan, 2020). However, the boundaries between such organizations and consulting are not always clear or strong.

One important and well-established factor in the use and valuing of knowledge is whether the source is seen as *internal* or *external* to the user or client. This has social and economic dimensions reflected in the notion of the so-called "not invented here" syndrome where knowledge is rejected or under-valued solely on the basis of its *external* origin or the opposite, where externally generated knowledge is seen as having greater credibility (Menon and Pfeffer, 2003). Likewise, studies of outsourcing use Transaction Cost Economics to establish rational motives and means to assess the relative benefits of "make or buy" with respect to services, including consultancy (Armbrüster, 2006). What this means in the context of seeking to support alternatives to external consultancy is recognizing the limitations of externally provided knowledge sources and considering how internal resources could be developed and when they are most appropriate. As noted above, this is evident in the typical guidelines for external consultancy purchasing decisions, in terms of using existing line management or specialist internal consulting or policy units. The latter have received much less research attention than external consultancy, but are in widespread use in both public and private sectors although they vary in their title, form, and structural position in the organization. In general, they have lacked the status (and marketing activity) of external consultancy and so, aside from resourcing expertise internally, action is needed in generating greater credibility for internal knowledge, in line with NIH. This might also allow for a greater legitimacy around alternatives to management as a form of knowledge for organizing, including established practices associated with a civil service ethos or national traditions of policy making.

While some research studies and policy guidelines consider or implicate alternative providers of organizing knowledge, the role of actors beyond clients/purchasers and consultants in governing or monitoring external management consultancy is rarely considered. Four particular groups should be recognized and, possibly, strengthened. Firstly, while the informal gatekeeper role of some internal client actors such as former external consultants has been identified (Sturdy and Wright, 2011), we have little knowledge of a form of consultancy directed at evaluating or auditing consultancy decisions and use—*third party or "meta" consulting* (Mohe, 2007). They are seen to be especially useful in conditions of conflicting interests between different client actors as well as between clients and consultants. Despite the obvious irony of using consultants to assess consultants, much like "independent advice" in the financial world, meta consultants cannot be eligible for other consulting work. Similarly, there are many options as to how they could be organized—as a "single, government-controlled organization, a private monopolist, a small group of companies competing in an oligopolistic market, or a large number of small service providers" (Stahl, 2018, 134). In the former case for example, the role of national audit offices (which report to parliaments rather than governments) could be expanded significantly. This resonates with calls for state run auditing for large firms (Brooks, 2019) and for the increasing relevance of a role for government in the regulation of business practices more generally (Kourula et al., 2019). Such a development could serve as a challenge to the ideology of self-regulation (Robson et al., 1994); although, so long as government remains a significant *consumer* of consultancy services and partner of consulting firms, it may remain disinclined to act (Muzio et al., 2011a). At the same time, there is also a potential to strengthen existing internal audit functions in organizations and expand their jurisdiction further into consulting use (Christopher, 2019).

Other monitors are more familiar to us. For example, it is widely recognized in studies of accountability and regulation that the *media* can play a vital role. However, with the decline of legacy media, this is under threat (Wedel, 2014) and, as we have noted, the attention can lead to secrecy, defensiveness, and impervious reputations. Nevertheless, such a role has been evident in the field of consultancy, including in the development of purchasing regulations (Pemer and Skjølsvik, 2018), but also in drawing attention to malpractice and "excess." Often, it first emerges at the margins, from investigative journalists along with activist individuals or organizations such as Spinwatch, *Private Eye*, Greenpeace, trade unions (e.g. Greenpeace, 2011; Unison, 2003) and other forms of "stakeholder activism" (Cundhill et al., 2018). While there is sometimes a danger of scapegoating consultants for wider problems with management knowledge and capitalism, such groups warrant more support as they are often sidelined in the lobbying process (Cave and Rowell, 2014). At the same time, academics have a role to play. To date, they have helped reveal some of the dangers and costs of consulting use, especially in political, policy and critical management studies, but activist scholarship remains marginal and cannot rely simply on calls for evidence-based management and policy (Fleming, 2019; Reay et al., 2009). Rather, shaping governance implies engaging in the politics of drawing or moving the line between what is deemed right and wrong (Palmer, 2012).

Conclusion

It is now widely established that the use of management consultancy in public policy and other contexts can have negative outcomes for democracy and organizational efficiency and effectiveness. These are based on a range of factors, but consultancy's ambiguity, commercial form, weak regulation, growth, and power imbalances combined with the inevitability of some need for external advice make effective governance an urgent matter. And yet, unlike in other occupational domains, there has been little, if any, attempt to explore and review practices and possibilities, especially outside of the field of purchasing. In this chapter, we have examined existing forms of governance and some of their limitations in terms of not being implemented, implementable, or sufficient to reflect the full extent of challenges consultancy use presents, particularly in complex and congested policy environments and where large PSFs can dominate agendas. In response, a number of emergent possibilities were outlined, beyond simply reinforcing existing mechanisms. In particular, attention was given to hybrid or contingent approaches to various dilemmas presented by governance and purchasing; to diverse providers, monitors and organizational forms; and to the fact that the problem of *management consultancy* includes issues surrounding both the processes of paid consultancy, whatever the content, *and* management, as a form of knowledge and practice (see Table 1).

Together, this framework represents a broad base from which to consider reforming the use and practice of management consulting, but it is by no means exhaustive and new challenges and possibilities are emerging. Furthermore, given the relative neglect of consulting governance in academic studies, there is considerable scope for further research.

Table 1 Governance options for the use of management consultancy

Existing governance	*Emergent prescriptions*
• Indirect regulation (e.g., auditing) • State certification (limited) • Professional/organizational codes of conduct and standards • Purchasing regulation, guidelines and professional practices (monitoring, internalizing, and contracting) • "Business appointment rules" (Targeting revolving doors) • Choice of alternatives—bans/budgetary limits (reputation)	• Reinforcing existing governance • Hybrid regimes of accountability, purchasing, and ethics (e.g., conditional accountability; from control to incentive-based measures of purchasing; whistleblowing; "system stewardship"; de-individualized ethics) • Alternative actors, structures, and knowledge (e.g., diverse consulting forms, knowledges, and reward structures; serial purchasing; alternative and credible internal and external providers; monitoring by the state, activists, media, academics, and meta consultancy)

We now take each of these—new possibilities and further research—briefly in turn. Firstly, the context of consultancy is changing, above and beyond issues already discussed, and this will impact governance. For example, *new paradigms of innovation* are emerging which, informed by developments such as climate change and the financial crisis, challenge assumptions of progress and economic growth from organizational change (Schot and Steinmueller, 2018). This could help legitimize alternative forms of knowledge in the field of organizing and public administration. Likewise, new approaches of *"tentative" governance* seek to address ambiguity and fluidity in innovation which matches some of that associated with consultancy, in transnational spaces for example (Kuhlmann, 2018). At the same time, technological developments are being directed toward improving public engagement (Hendricks and Carson, 2008) and to "*smart governance*." Indeed, consulting firms are active in promoting these approaches, but have not had the spotlight turned back on themselves. For example, with the rise of internet-based interactions, fuelled by Covid19 working, client-consultant relations could become more readily monitored. Likewise, it seems plausible that sharing information about consulting providers is easier now with internet-based technologies—ratemyconsultancy.com. Indeed, management ideas are now being produced and debated more interactively on the internet compared to when they were controlled by a small number of publishers and large consulting firms (Barros and Rüling, 2019). This includes "thought leadership" which is now authored by different actors although the resources of large firms are still significant in dominating provision.

All these emerging developments warrant further research, but other, more fundamental issues need to be studied to help inform policy on the governance of consulting. Given the importance of its ambiguity and commercial pressures, comparative studies are needed on the nature and outcomes of different forms of consultancy, to include: internal and external provision; consulting in more or less ambiguous and contested knowledge domains (e.g. engineering vs arts-based initiatives); consulting and other forms of knowledge providers, including in non-professional contexts. For example, while the potential for negative outcomes from management consultancy use is relatively well established, we do not know whether these risks are greater or less with other channels such as with thinktanks or university policy labs. Likewise, the relative legitimacy or credibility of knowledge sources in different national and policy contexts is not well understood along with ways in which it can be increased (Sahir and Brutus, 2018). Similarly, although it seems clear that existing governance is neither fully effective nor sufficient, the reasons behind this have yet to be firmly established—resources, skills, commitment. What are the barriers to governments regulating consultancy use or to large firms engaging with professionalization?

At a more general level, it is clear that there are no "magic bullets" of reform to address the problems of consultocracy or of consulting "over-use" more generally. Nevertheless, addressing the problems arising from its ambiguity and commercial relations as well as from the particular form of knowledge typically promoted in consulting is an obvious starting point. Some caution is required given the scant nature of data on consulting and its influence, but this chapter has begun to outline challenges, possibilities, and emerging opportunities—other options are available.

Acknowledgements Thanks to the following for comments on an earlier draft of this text: Ian Kirkpatrick; Svenja Keele, Terry McNulty, Jost Sieweke, Andreas Werr and the editors, Chris Hurl and Anne Vogelpohl.

References

Armbrüster, T. (2006) *The Economics and Sociology of Management Consulting*, Cambridge: Cambridge University Press.

Ayers, S. and Kaplan, S.E. (2005) 'Wrongdoing by consultants: An examination of employees' reporting intentions', *Journal of Business Ethics* 57: 121–137.

Barros, M. and Rüling, C.-C. (2019) 'Business media: From gatekeeping to transmediality', in A.J. Sturdy, S. Heusinkveld, T. Reay and D. Strang (eds.) *The Oxford Handbook of Management Ideas*, Oxford: Oxford University Press, pp. 195–215.

Boussebaa, M. and Faulconbridge, J.R. (2019) 'Professional service firms as agents of economic globalization: A political perspective', *Journal of Professions and Organization* 6: 72–90.

Brint, S. (1990) 'Re-thinking the policy influence of experts', *Sociological Forum* 5(3): 361–385.

Brooks, R. (2019) *Beancounters*, London: Atlantic Books.

Cave, T. and Rowell, A. (2014) *A Quiet Word—Crony Capitalism and Broken Politics in Britain*, London: The Bodley Head.

Christensen, C.M., Wang, D. and van Bever, D. (2013) 'Consulting on the cusp of disruption', *Harvard Business Review* 2013(10): 106–115.

Christopher, J. (2019) 'The failure of internal audit', *Journal of Management Inquiry* 28(4): 472–483.

Collins, D. and Butler, N. (2020) 'Success and failure in professional projects: The nature, contours and limits of consulting professionalism', *British Journal of Management* 31(3): 457–469.

Corporate Reform Collective (CRC) (2014) *Fighting Corporate Abuse*, London: Pluto.

Craft, J. and Halligan, J. (2020) *Advising Governments in the Westminster Tradition: Policy Advisory Systems in Australia, Britain, Canada and New Zealand.* Cambridge: CUP.

Craft, J. and Howlett, M. (2013) 'The dual dynamics of policy advisory systems: The impact of externalization and politicization on policy advice', *Policy and Society* 32(3): 187–197.

Cundhill, G.J., Smart, P. and Wilson, H.N. (2018) 'Non-financial shareholder activism: A process model for influencing corporate environmental and social performance', *International Journal of Management Reviews* 20(2): 606–626.

David, R.J., Sine, W.D. and Haveman H.A. (2013) 'Seizing opportunity in emerging fields: How institutional entrepreneurs legitimated the professional form of management consulting', *Organization Science* 24: 356–377.

Engwall, L., Kipping, M. and Usdiken, B. (2016) *Defining Management: Business Schools, Consultants, Media*, London: Routledge.

Faust, M. and Schneider, K. (2014) 'Functional equivalents to external consulting—A case of a reluctant German corporation', in M. Faust (ed.) *Globale Managementberatung*, Berlin: Rainer Hampp Verlag, pp. 155–179.

FEACO (2018) *Survey of European Management Consultancy 2017-2018*, Brussels: FEACO.

Financial Times (2019) 'Watchdog's plans for audit reform get mixed reception', *Financial Times*, 18 April, https://www.ft.com/content/40f12080-61d6-11e9-a27a-fdd51850994c.

Fleming, P. (2019) 'Robots and organization studies: Why robots might not want to steal your job', *Organization Studies* 40(1): 23–38.

Furusten, S., Werr, A., Ardenfors, M. and Walter, S. (2013) 'CSR and the consultancy sector', in M. Jutterström and P. Norberg (eds.) *CSR as a Management Idea—Ethics in Action*, Cheltenham: Elgar, pp. 75–93.

Freidson, E. (2001) *Professionalism, the Third Logic*, Chicago: University of Chicago Press.

Gibbons, A. (2019) *The accountability of management consultancy use in the English NHS*. Paper presented at the IPPA Conference, 26–28 June, Montreal, Canada.

Glückler, J. and Armbrüster, T. (2003) 'Bridging uncertainty in management consulting: The mechanisms of trust and networked reputation', *Organization Studies* 24(2): 269–297.

Greenpeace (2011) *Bad Influence: How McKinsey Plans Lead to Rainforest Destruction*, Greenpeace International, Amsterdam.

Greenwood, R. and Empson, L. (2003) 'The professional partnership—Relic or exemplary forms of governance?', *Organization Studies* 24(6): 909–933.

Gross, C. and Kieser, A. (2006) 'Are consultants moving towards professionalization?' *Research in the Sociology of Organizations* 24: 69–100.

Grundmann, R. (2018) 'The rightful place of expertise', *Social Epistemology* 32(6): 372–386.

Guerin, B., McCrae, J. and Shepheard, M. (2018) *Accountability in modern government—Recommendations for change*, London: Institute for Government.

Haines, F (2013) 'Three risks, one solution? Exploring the relationship between risk and regulation', *The Annals of the American Academy of Political and Social Science* 649(1): 35–51.

Harvey, W.S., Morris, T. and Müller Santos M. (2017) 'Reputation and identity conflict in management consulting', *Human Relations* 70(1): 92–118.

Hendricks, C.M. and Carson, L. (2008) 'Can the market help the forum? Negotiating the commercialization of deliberative democracy', *Policy Sciences* 41: 293–313.

Heusinkveld, S., Sturdy, A.J. and Werr, A. (2011) 'The co-consumption of management ideas and practices', *Management Learning* 42(2): 1–9.

Hodge, G. and Bowman, D. (2006) 'The "consultocracy": The business of reforming government', in G. Hodge (ed.) *Privatization and Market Development: Global Movements in Public Policy Ideas*, Cheltenham: Edward Elgar, pp. 97–126.

Honer, D. and Mohe, M. (2009) 'Behind clients' closed doors: What hinders client firms from professionally dealing with consultancy?', *Scandinavian Journal of Management* 25: 299–312.

House of Commons (2019) *The future of audit*, London: HMSO.

ISO (2017) https://www.iso.org/news/ref2160.html, accessed 27 October 2020.

Institute for Government (2019) *Government outsourcing—What has worked and what needs reform?* London: IFG.

Jones, C.W. (2019) 'Adviser to the king—Experts, rationalization and legitimacy', *World Politics* 71(1): 1–43.

Karpoff, J.M. (2012) 'Does reputation work to discipline corporate misconduct?' in T.G. Pollock and M.L. Barnett (eds.) *The Oxford Handbook of Corporate Reputation*, Oxford: OUP, pp. 361–383.

Keele, S. (2018) *Outsourcing adaptation: Examining the role and influence of consultants in governing climate change adaptation*, University of Melbourne, Melbourne.

Kipping, M. (2011) 'Hollow from the start? Image professionalism in management consulting', *Current Sociology* 59: 530–550.

Kipping, M. and Kirkpatrick, I. (2012) 'Alternative pathways of change in professional services firms: The case of management consulting', *Journal of Management Studies* 50(5): 777–807.

Kipping, M. and Saint-Martin, D. (2005) 'Between regulation, promotion and consumption: Government and management consultancy in Britain', *Business History* 47(3): 447–465.

Kipping, M., Buhlmann, F. and David, T. (2019) 'Professionalization through symbolic and social capital: Evidence from the careers of elite consultants', *Journal of Professions and Organization* 6(3): 265–285.

Kirkpatrick, I., Sturdy, A.J., Veronesi, G., et al. (2018) 'The impact of management consultants on public service efficiency', *Policy & Politics* 47(1): 77–96.

Kourula, A., Moon, J., Salles-Djelic, M.-L. and Wickert, C. (2019) 'New roles of government in the governance of business conduct', *Organization Studies* 8: 1101–1124.

Kuhlmann, S. (2018) 'Introduction to discussion paper on 'three frames for innovation policy: R&D, systems of innovation and transformative change'', *Research Policy* 47: 1553.

Lindberg, K. and Furusten, S. (2005) 'Breaking laws: Making deals', in S. Furusten and A. Werr (eds.) *Dealing with Confidence*, Copenhagen: CBS Press, pp. 169–183.

Lonsdale, C., Hoque, K., Kirkpatrick, I. and Sanderson, J (2017) 'Knowing the price of everything? Exploring the impact of increased procurement professional involvement on management consultancy purchasing', *Industrial Marketing Management* 65: 157–167.

Marsh, S. (2009) *The Feminine in Management Consulting—Power, Emotion and Values in Consulting Interactions*, Basingstoke: Palgrave.

Martin, J.F. (1998) *Reorienting A Nation: Consultants and Australian Public Policy*, Aldershot: Ashgate.

McKenna, C.D. (2006) *The World's Newest Profession: Management Consulting in the Twentieth Century*, Cambridge: Cambridge University Press.

McKenna, C. (2007) 'Give professionalization a chance', in D. Muzio, S. Ackroyd and J.F. Chanlat (eds.) *Redirections in the Study of Expert Labour: Established Professions and New Expert Occupations*, Houndmills: Palgrave Macmillan.

Meere, D. (2019) 'Consultancy industry faces its own challenger bank moment'. *Consultancy.UK*, 23 July.

Menon, T. and Pfeffer, J. (2003) 'Valuing internal vs. external knowledge: Explaining the preference for outsiders', *Management Science* 49(4): 497–513.

Mitchell, A. and Sikka, P. (2011) *The pin-striped mafia—How accountancy firms destroy societies*, Basildon, Essex: Association for Accountancy and Business Affairs.

Mohe, M. (2005) 'Generic strategies for managing consultants: Insights from clients' companies in Germany', *Journal of Change Management* 5(3): 357–365.

Mohe, M. (2006) 'Daimler Chrysler's Consulting Infobase', *Consulting to Management* (no further details) p. 32.

Mohe, M. (2007) 'Meta-consulting: Idea, Functions and Limitations of a New Business Model in the Consulting Market'. Workshop on "Innovative Capabilities and the Role of Consultants in the Information Economy" ZEW, 19–20 November, Mannheim, Germany.

Morgan, G., Sturdy, A.J. and Frenkel, M. (2019) 'The role of large management consultancy firms in global public policy', in D. Stone and K. Moloney (eds.) *The Oxford Handbook of Global Policy and Transnational Administration*, Oxford: Oxford University Press.

Muzio, D., Faulconbridge, J., Gabbioneta, C. and Greenwood, R. (2016) 'Bad apples, bad barrels and bad cellars: A 'boundaries' perspective on professional misconduct', in D. Palmer, R. Greenwood and K. Smith-Crowe (eds.)

Organizational Wrongdoing, Cambridge, UK: Cambridge University Press, pp. 141–175.

Muzio, D., Kirkpatrick, I. and Kipping, M. (2011a) 'Professions, organizations and the state: Applying the sociology of the professions to the case of management consultancy', *Current Sociology* 59: 805–824.

Muzio, D., Hodgson, D., Faulconbridge, J., Beaverstock, J. and Hall, S. (2011b) 'Towards corporate professionalization: The case of project management, management consultancy and executive search', *Current Sociology* 59: 443–464.

NAO (National Audit Office) (2006) *Central Government's Use of Consultants*, London: HMSO

NAO (2016) *Cross-Government Use of Consultants and Temporary Staff*, London: HMSO.

NAO (2017) *Investigation into Government's Management of the Business Appointment Rules*, London: National Audit Office.

Protess, B. and Silver-Greenberg, J. (2013) 'US Efforts to regulate consultants face big obstacles', *New York Times*, 10 April.

OECD (2009) *Principles for enhancing integrity in public procurement*, Paris: OECD.

OECD (2017) *Policy advisory systems: Supporting good governance and sound public decision making*, Paris: OECD.

O'Flynn, J. and Sturgess, G.L. (2019) *2030 and beyond—Getting the work of government done*. ANZSOG.edu.au.

O'Mahoney, J. (2011) 'Advisory anxieties: Ethical individualisation in the UK consulting industry', *Journal of Business Ethics* 104(1): 101–113.

O'Mahoney, J., Heusinkveld, S. and Wright, C. (2013) 'Commodifying the commodifiers: The impact of procurement on management knowledge', *Journal of Management Studies* 50(2): 204–235.

O'Mahoney, J. and Sturdy, A.J. (2016) 'Power and the diffusion of management ideas: The case of McKinsey & Co.', *Management Learning* 47(3): 247–265.

Owen, D. (2019) 'Consultants, the environment and the law', *Arizona Law Review* 61(4): 824–884.

Parker, M. (2018) *Shut Down the Business School*, London: Pluto Press.

Palmer, D (2012) *Normal Organizational Wrongdoing*, Oxford: Oxford University Press.

Palmer, D., Smith-Cowe, K. and Greenwood, R. (2016) 'The imbalances and limitations of theory and research on organizational wrongdoing', in D. Palmer, K. Smith-Crowe and R. Greenwood (eds.) *Organizational Wrongdoing*, Cambridge, UK: Cambridge University Press, pp. 1–6.

Patrucco, A.S., Walker, H., Luzzini, D. and Ronchi, S. (2018) 'Which shape fits best? Designing the organisational form of local government procurement', *JOPSM* 25(3): 100504.

Pemer, F., Sieweke, J. and Werr, A. et al. (2014) 'The cultural embeddedness of professional service purchasing', *Journal of Purchasing and Supply Management* 20(4): 273–285.
Pemer, F., Werr, A. and Börjeson, L. (2020) 'The role of chief executive tenure for public organizations' hiring of management consultants', *Governance* 33(2), 258–269.
Pemer, F. and Skjølsvik, T. (2018) 'Adopt or adapt? Unpacking the role of institutional work processes in the implementation of new regulations', *Journal of Public Administration Research and Theory* 28(1): 138–154.
Ramanna, K. (2019) *Building a culture of challenge in audit firms.* Paper commissioned by PwC UK.
Raudla, R. (2013) 'Pitfalls of contracting for policy advice: Preparing performance budgeting reform in Estonia', *Governance-an International Journal of Policy Administration and Institutions* 26(4): 605–629.
Reay, T., Berta, W. and Kohn, M. (2009) 'What's the evidence on evidence-based management?', *Academy of Management Perspectives* 23(4): 5–8.
Richardson, J. (2012) 'New governance or old governance?' in D. Levi-Faur (ed.) *The Oxford Handbook of Governance*, Oxford: OUP.
Robson, K., Willmott, H., Cooper, D. and Puxty, T. (1994) 'The ideology of professional regulation and the markets for accounting labour', *Accounting Organizations and Society* 19(6): 527–523.
Roodhooft, F. and van den Abbeele, A. (2006) 'Public procurement of consulting services', *International Journal of Public Sector Management* 19(5): 490–512.
Sahir, R. and Brutus, S. (2018) 'A view of the role of experts in corporate consulting', *Consulting Psychology Journal* 70(2): 95–112.
Saint-Martin, D. (2000) *Building the New Managerialist State: Consultants and the Politics of Public Sector Reform in Comparative Perspective*, Oxford: Oxford University Press.
Saint-Martin, D. (2012) 'Governments and management consultants: Supply, demand and effectiveness', in T. Clark and M. Kipping (eds.) *The Oxford Handbook of Management Consulting*, Oxford: OUP, pp. 447–464.
Schot, J. and Steinmueller, W.E. (2018) *Three frames for innovation policy: R&D, systems of innovation and transformative change.* Research Policy no. 47.
Shaw, D. (2019) 'Aristotle and the management consultants: Shooting for ethical practice', *Philosophy of Management* 19: 21–44.
Sieweke, J., Birkner, S. and Mohe, M. (2012) 'Preferred supplier programmes for consulting services', *Journal of Purchasing and Supply Management* 18(3): 123–136
Stahl, J. (2018) 'Impact of management consulting—Perspectives on value creation, performance evaluation, and governance of the client-consultant relationship', PhD thesis, Zeppelin University, Konstanz.

Stapper, E.W., Van der Veen, M. and Janssen-Jansen, L.B. (2020) 'Consultants as intermediaries: Their perceptions on citizen involvement in urban development', *Environment and Planning C: Politics and Space* 38(1): 60–78.

Steiner, R., Kaiser, C. and Reichmuth, L. (2018) 'Consulting for the public sector in Europe', in E. Ongaro and S. van Thiel (eds.) *The Palgrave Handbook of Public Administration and Management in Europe*, London: Palgrave Macmillan, pp. 475–495.

Sturdy, A.J. (2009) 'Popular consultancy critiques and a politics of management learning?' *Management Learning* 40(4): 457–463.

Sturdy, A. and O'Mahoney, J. (2018) 'Explaining national variation in the use of management consulting knowledge: A framework', *Management Learning* 49(5): 537–558.

Sturdy, A.J. and Wright, C. (2011) 'The active client: The boundary-spanning roles of internal consultants as gatekeepers, brokers and partners of their external counterparts', *Management Learning* 42(5): 485–503.

Sturdy, A.J., Handley, K., Clark, T. and Fincham, R. (2009) *Management Consultancy, Boundaries and Knowledge in Action*, Oxford: Oxford University Press.

Taminiau, Y. and Heusinkveld, S. (2019) *Inter-Professional Dynamics Within Complex Organisations*. Amsterdam, Netherlands: VU Amsterdam. Working paper.

Tyllstrom, A (2019) 'More than a revolving door—Corporate lobbying and the socialisation of institutional carriers', *Organization Studies*, https://doi.org/10.1177/0170840619848014.

Unison (2003) *Stitched Up: How the Big Four Accountancy Firms Have PFI Under Their Thumbs*, London: Unison.

Villena, V.H., Revilla, E. and Choi, T.Y. (2011) 'The dark side of buyer-supplier relationships—A social capital perspective', *Journal of Operations Management* 29: 561–576.

Vogelpohl, A. (2018) 'Consulting as a threat to local democracy? Flexible management consultants, pacified citizens, and political tactics of strategic development in German cities', *Urban Geography* 1–21.

Vriens, D., Vosselman, E. and Gross, C. (2018) 'Public professional accountability—A conditional approach', *Journal of Business Ethics* 153(4): 1179–1196.

Wedel, J. (2014) *Unaccountable—How Elite Power Brokers Corrupt Our Finances, Freedom, and Security*, New York: Pegasus.

Weiss, A.E. (2019) *Management Consultancy and the British State*, Basingstoke: Palgrave.

Werr, A. and Pemer, F. (2007) 'Purchasing management consulting services: From management autonomy to purchasing involvement', *Journal of Purchasing and Supply Management* 13(2): 98–112.

Werr, A. and Styhre, A. (2002) 'Management consultants-friend or foe? Understanding the ambiguous client-consultant relationship', *International Studies of Management & Organization* 32(4): 43–66.

Wright, C. and Kitay, J. (2002) 'But does it work? Perceptions of the impact of management consulting', *Strategic Change* 11(5): 271–278.

Ylönen, M. and Kuusela, H. (2019) 'Consultocracy and its discontents: A critical typology and a call for a research agenda', *Governance* 32(2): 241–258.

The manufacturer's authorised representative in the EU is Springer Nature Customer Service Centre GmbH, Europaplatz 3, 69115 Heidelberg, Germany. If you have any concerns regarding our products, please contact ProductSafety@springernature.com

Printed and bound by CPI Group (UK) Ltd, Croydon, CR0 4YY

30/09/2026

02225451-0016